This edition published in 1993 by Mimosa Books, distributed by Outlet Book Company, Inc., a Random House Company, 40 Engelhard Avenue, Avenel, New Jersey 07001.

10 9 8 7 6 5 4 3 2 1

First published in 1993 by Grisewood & Dempsey Ltd.

ISBN 1 85698 502 4

Printed and bound in Italy

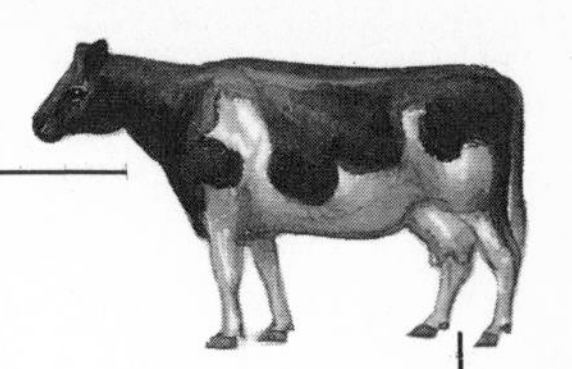

THE COW

By Angela Royston
Illustrated by Bob Bampton

MIMOSA
·BOOKS·

NEW YORK • AVENEL, NEW JERSEY

The young cows have spent all winter in the barn but now it is spring and the farmer brings them out into the fields. They smell the fresh grass and pull up huge mouthfuls of it. They run across the field in the bright sunshine, happy to be outside again.

The black cow sees that the gate to the next field has been left open. She pushes through it and the other cows follow her. This field is full of long grass and clover. The cows love the damp sweet smells and they eat all they can.

The farmer knows that clover will make the cows ill so he drives them back to their own field. But later that day the cow's aching stomach swells up with gas. She is so uncomfortable she cannot lie down.

The other cows are ill too, so that evening the vet comes. She decides to make a small hole in the black cow's belly to let the gas out, but the other cows just need medicine. Next day they are all well again.

Summer comes, and the cows are nearly two years old. They are grazing alongside the riverbank. The black cow pulls the grass into her mouth with her long tongue. Flies buzz around her and she flicks her tail to drive them away.

The leader of the herd moves away from the riverside and the others follow her. They stand in the shade of a big tree and chew their cud. They bring some of the grass back into their mouths, chew it slowly, and then swallow it again.

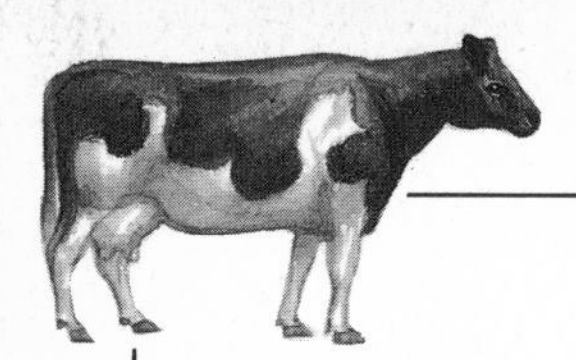

The next day the farmer puts a bull into the field. The black cow is curious and wanders over to have a better look at him. They sniff each other and stay close together all day. The bull paws the ground and nuzzles her side. He knows by her smell that she is ready to mate.

The days pass and winter comes. The nights are very cold and there is less grass to eat in the field. One rainy afternoon as the cows huddle together, the farmer comes to take them back to the barn.

The floor is covered with straw and there is plenty of hay to eat. The black cow tosses the straw with her head and kicks it up with her heels. She hears the farm dog barking and a tractor rumbling by.

The black cow stays in the barn all winter. During the day the hens peck in the straw for grain, and at night the mice sometimes come looking for grain too. A new calf is growing inside the cow and as winter turns to spring her belly grows larger. Ten months after mating, her calf is ready to be born.

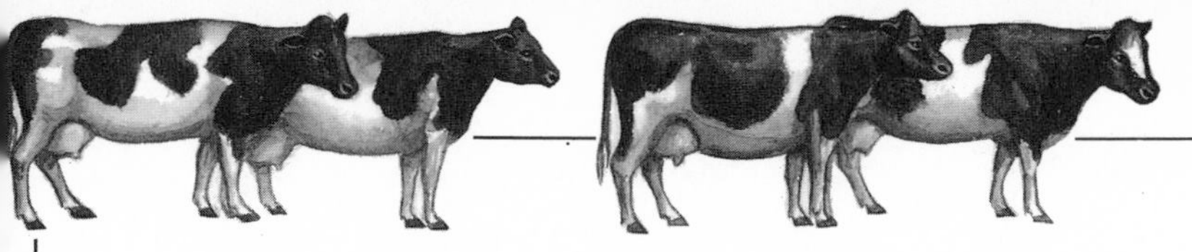

The farmer puts her in a quiet pen with lots of straw. She lies down, and a few hours later her calf is born. She licks him all over.

The black cow nuzzles her newborn calf and helps him to stand on his wobbly legs. At first the calf falls down again but soon he is strong enough to reach her udders and suck milk from her teats.

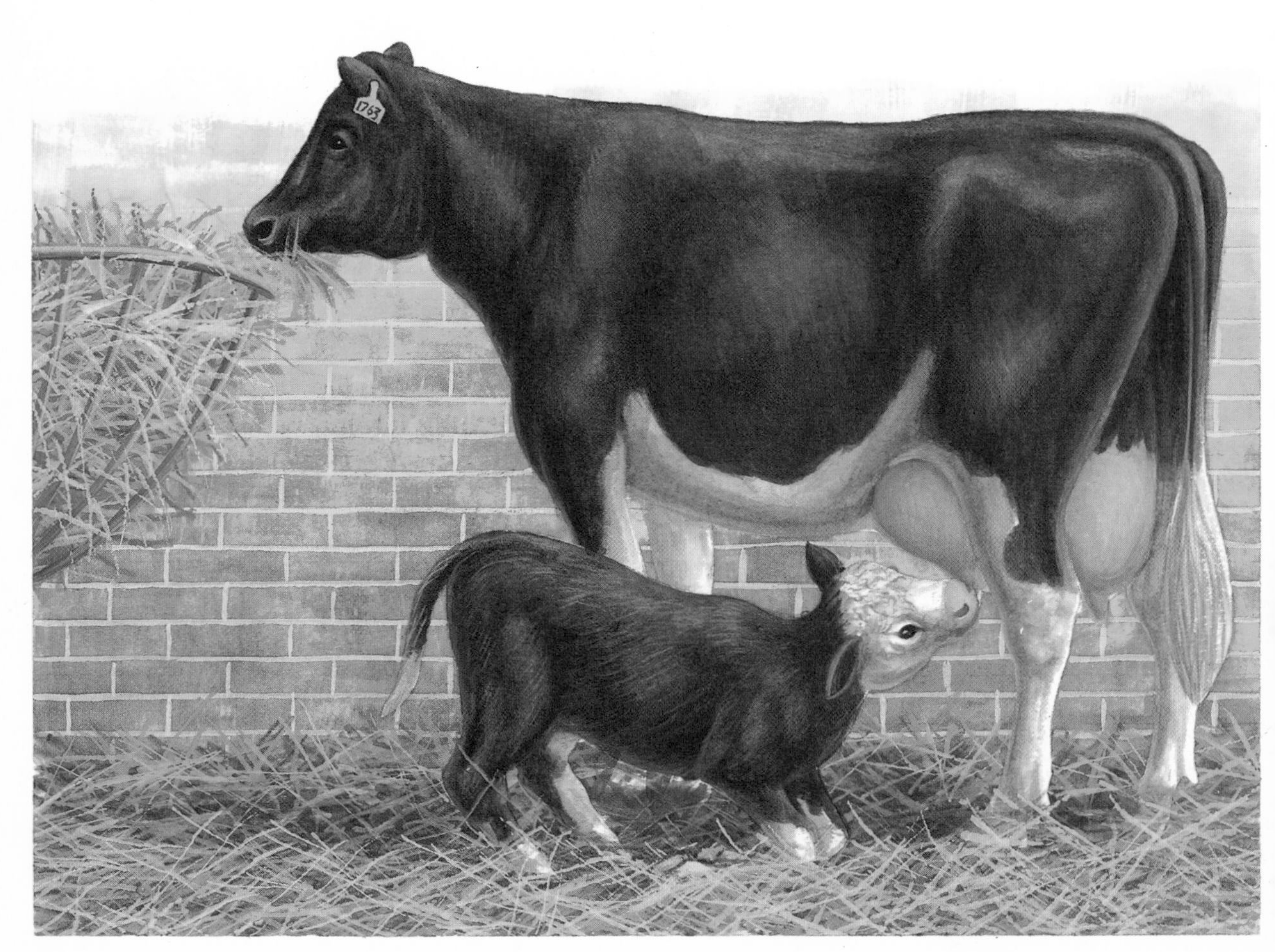

The cow can feel the little calf trembling against her side. She moos softly to him and they lie down together. The little calf is warm and tired and soon falls asleep.

When the calf is a few days old he is moved into a pen with other calves. He misses his mother but now he is old enough to suck milk from a teat attached to a bucket. The farmer takes his mother to be milked.

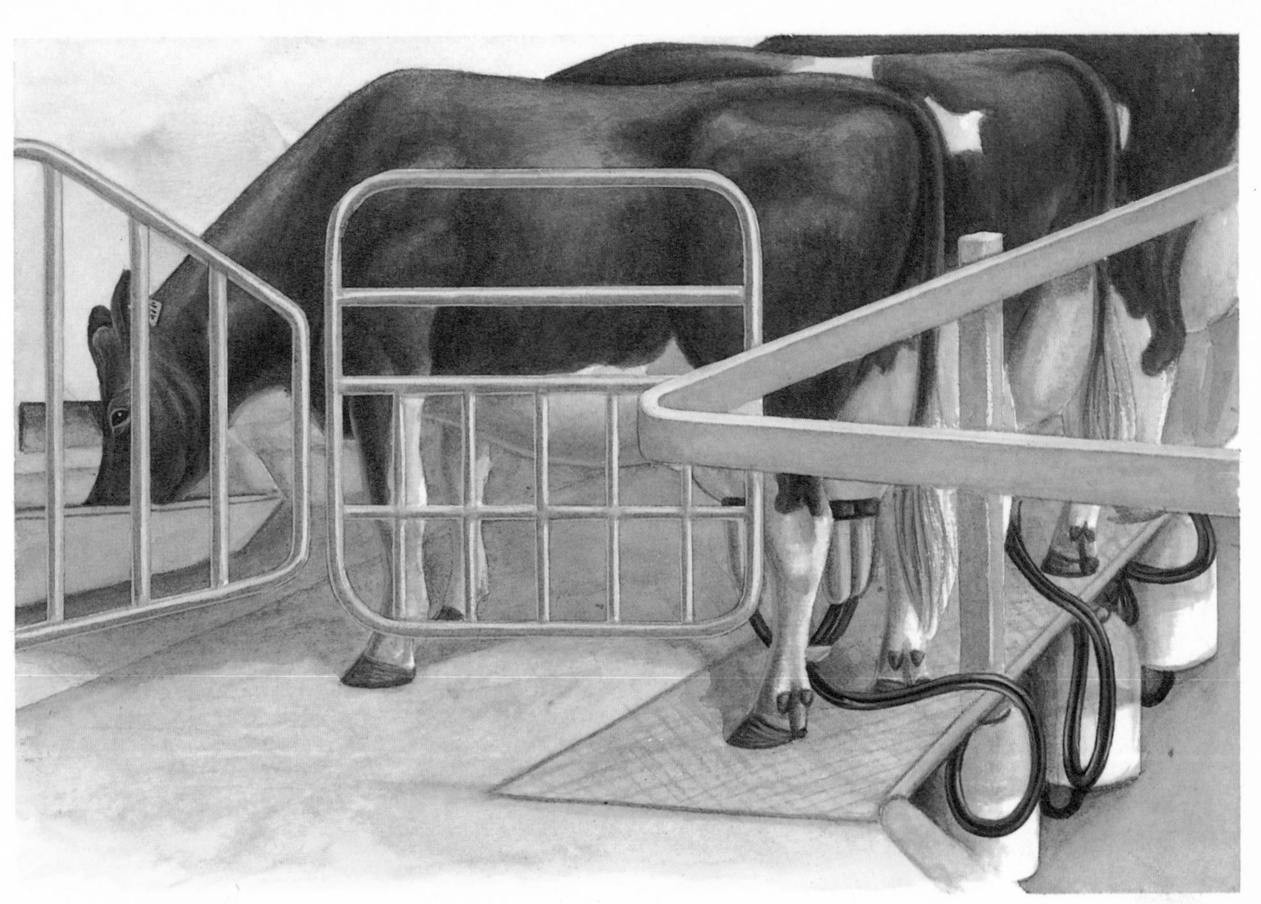

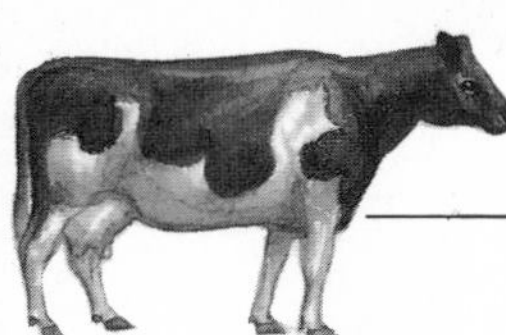

She stands in a stall and four rubber cups are placed over her teats. She munches feed from a trough and listens to the hum of machinery as the milk is gently pumped from her udders.

It is summertime again. The cow has not seen her calf since he was very young and now she has forgotten about him. It is late afternoon and her udders are full of milk. She waits by the gate because she knows it will soon be milking time.

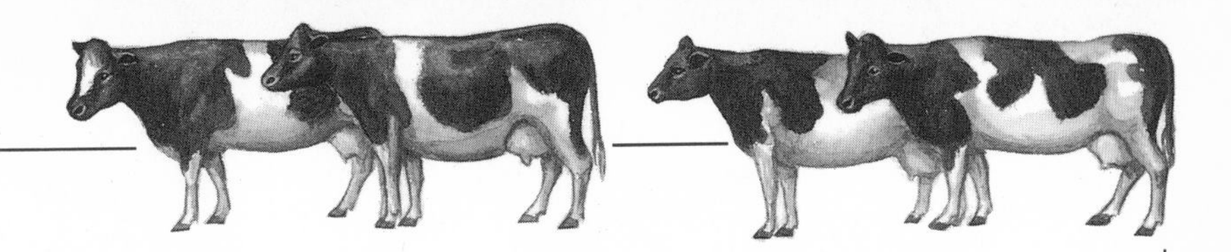

When the farmer comes she slowly follows the other cows up the lane to the milking parlor. A new calf is growing inside her. It will be born next spring.

More About Cows

There are many different types of dairy cow. The cow in this story is a Holstein cow. Holsteins and Friesians give the most milk of all dairy cows.

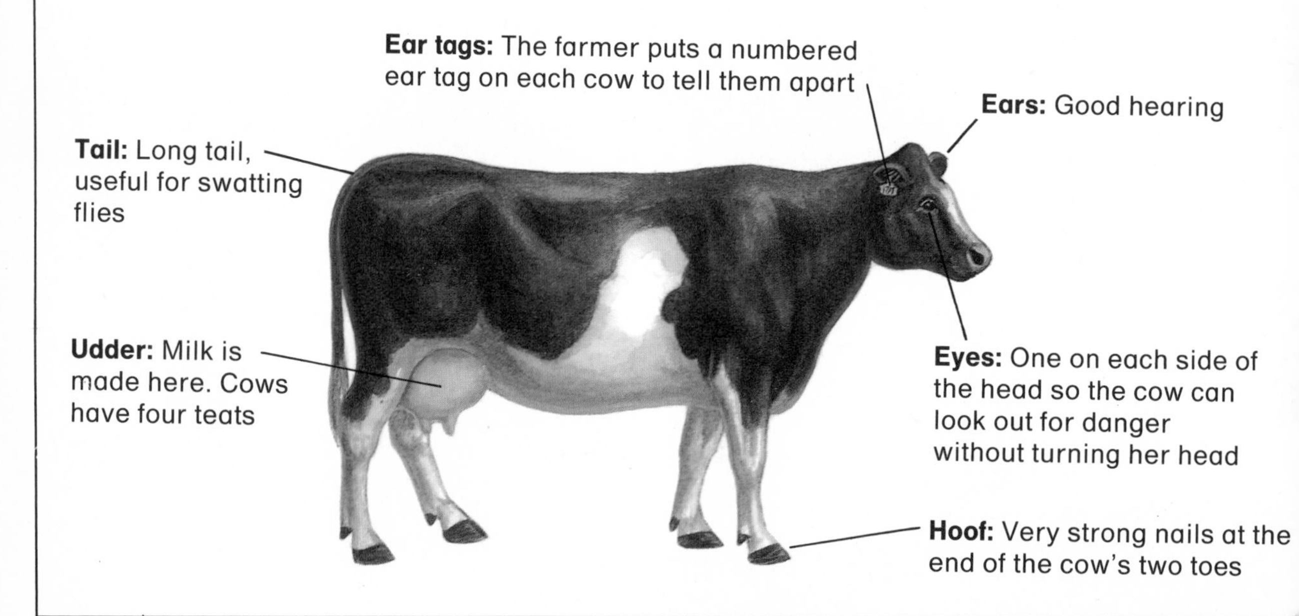

The Brown Swiss is a very old breed of cow. They used to pull plows and other farm machinery. Now they are popular in the high mountains of Switzerland. Jersey and Guernsey cows give very creamy milk. Milk can be used to make butter, yogurt and many kinds of cheese.

A cow has four stomachs. The first two store the food as she grazes. Later she brings balls or cuds of grass back into her mouth and chews them until they are mushy. She swallows them again and they then pass to the other stomachs to be digested.